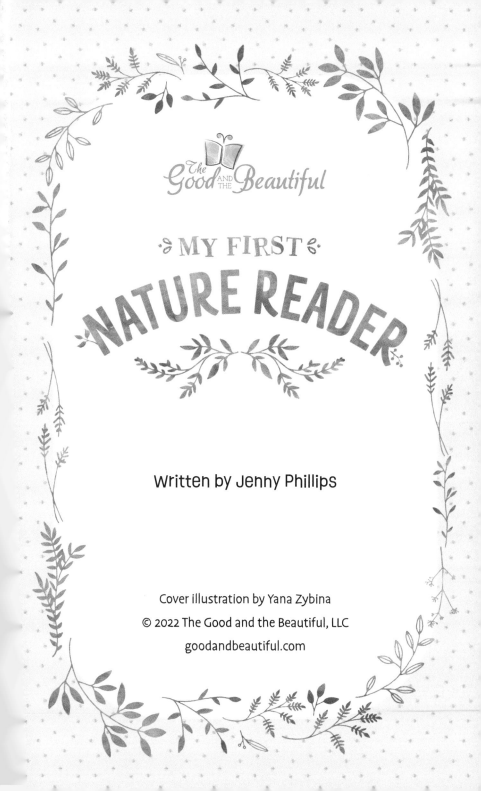

The Good AND THE Beautiful

MY FIRST
NATURE READER

Written by Jenny Phillips

Cover illustration by Yana Zybina
goodandbeautiful.com

TABLE OF CONTENTS

CAT

CVC Words

The Good and the Beautiful

A cat can nap.

It can get up.

Run! Run!

Sit in a lap.

Pat! Pat!

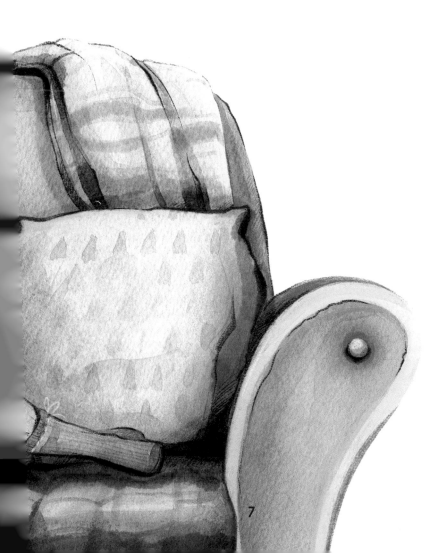

I pet the cat.

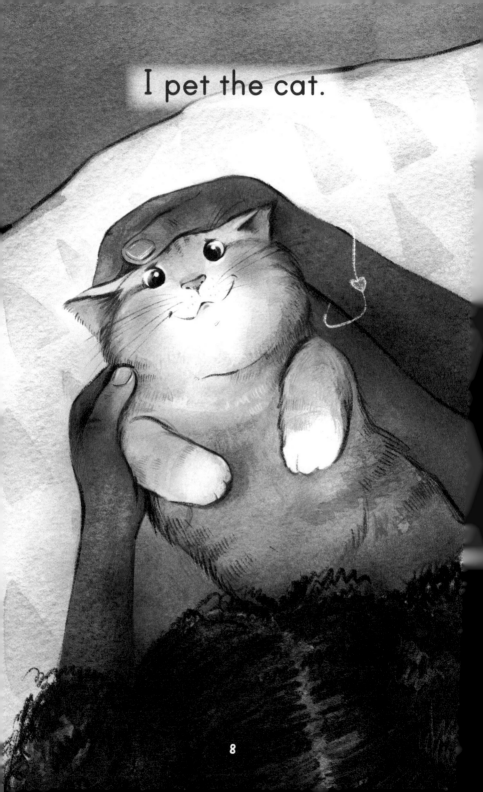

Book 2

DOG

CVC Words

The Good and the Beautiful

A dog can sit.

I let it tug.

Mud!

It can run.

It can get wet.

THE CUB

CVC Words

The Good and the Beautiful

I am a cub.

I nap.

I am up.

I dig in the mud.

I can fit in it.

I am wet.

Dan and
Dot love
the pig.

He can pet it.

She can hug it.

He can hum to it.

It can go in the mud.

"We love it!"

BOOK 5

BUGS

Words Where S Says /Z/

The Good and the Beautiful

I love bugs.

A bug is on the log.

A bug is on me.

A red bug has fun.

It can zig and zag.

Do you love bugs?

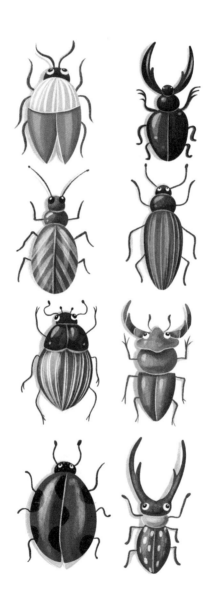

BOOK 6

THE FOX

CVC Words

The Good and the Beautiful

A fox can run.

It is not as big as a hog.

It is not as big as an ox.

It can get the rat.

Zip! It runs and runs.

It has six cubs.

It is a mom.

The fox is in the den.

BOOK 7

THE RAM

CVC Words

The Good and the Beautiful

The ram is big.

He loves to go on it.

He can go to the tip.

We love the ram.

He is at the top.

Run! Run, ram!

Book 8

THE WEB

Sight Words: Group 2

The Good and the Beautiful

We look for a web.

A web is on the log.

Webs are so fun.

A web is on me!

I do not love it.

69

Look! I do love the web.

THE SUN

CVC Words

The Good and the Beautiful

Is the sun up? No.

Is the sun up? Yes!

A lot of sun is on us.

I can sit in the sun.

Or I can run in the sun.

The sun can be so hot!

NUTS

Book 10

CK; ALL

Illustrated by Bojana Stojanovic

goodandbeautiful.com

The nut is up.

It can fall on the rock.

I pick it up.

I pop it.

Yum!

I get a sack
and call Dad.

He is tall.

It is fun to pick the nuts.

THE WIND

Ending Consonant Blends

The
Good and the Beautiful

The wind is so soft.

It can bend the crop.

It is as fast as it can be.

It can help the raft go.

Wind is the best!

BOOK
12

THE YAK

SS, FF, LL

The
Good AND THE Beautiful

The yaks are on the hill.

The yaks run past the mill.

Huff! Puff!

The yaks pass the pond.

Look! It has a bell.

The yaks rest on the moss.

The crab is on the sand.

The crab will get in.

He can swim.

He swims by the clam.

He will try to grab it.

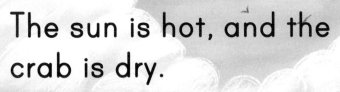

The sun is hot, and the
crab is dry.

It is two jays.

They love to play.

The hay is for
her nest.

There are two
gray eggs.

Oh! The little eggs crack!

A duck is in the pond.

Splash! A frog jumps in.

Flash! A fish swims past.

Swish! The plants sway.

126

The wind is fresh.

Hush! The duck will rest.

FINCH

CH

A finch has such fun.

He has his lunch.

It is a bunch of nuts.

He pecks at the moss,
and he gets a bug.

He sits on the branch.

Illustrated by Valentina Fedorova

goodandbeautiful.com

Seth loves moths.

The moths are on the cloth.

141

The moths flit and flap
on the path.

142

Seth is glad to be with them.

Book 18

THE SWING

ING

Illustrated by Valentina Fedorova

goodandbeautiful.com

It is spring! The jays sing.

I go up in my swing
and look at things.

Spring brings grass
and wings in the sky.

I love being
in spring!

Josh sees a bee by a plant.

He can spy the tree on a steep hill.

Josh plants seeds.

He pulls green
weeds.

Josh loves to see plants.

"Come," said Jenn, "let us spy some frogs."

Did they see any?
Yes! Many!

One, two, three frogs
jumped on a log.

What did the frogs do
then?

They hopped from the log.

"One frog landed in your bag," said Dad.

"What will you do?"

Jenn stepped to the bag.

She picked up the frog
and put it in the pond.

They grinned as they looked at the frogs.